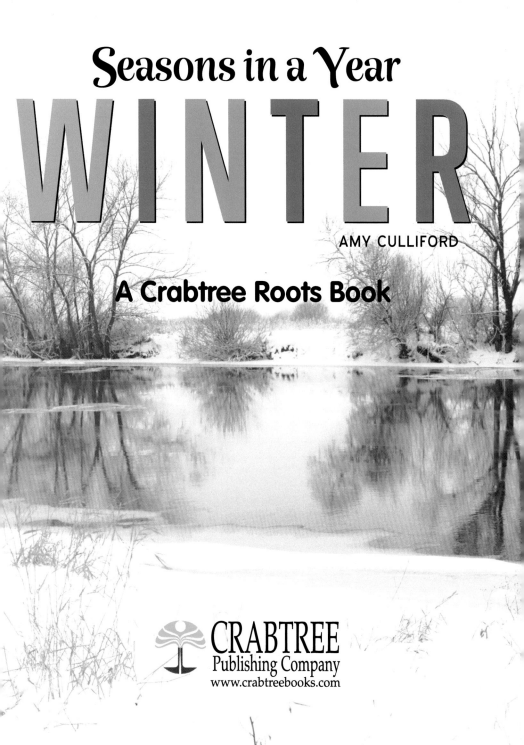

Seasons in a Year
WINTER

AMY CULLIFORD

A Crabtree Roots Book

CRABTREE
Publishing Company
www.crabtreebooks.com

School-to-Home Support for Caregivers and Teachers

This book helps children grow by letting them practice reading. Here are a few guiding questions to help the reader with building his or her comprehension skills. Possible answers appear here in red.

Before Reading:

• What do I think this book is about?
  - *This book is about a season called winter.*
  - *This book is about things you can see in winter.*

• What do I want to learn about this topic?
  - *I want to learn what people wear in winter.*
  - *I want to learn what winter looks like.*

During Reading:

• I wonder why...
  - *I wonder why winter is cold.*
  - *I wonder why people wear hats.*

• What have I learned so far?
  - *I have learned that people like to skate in winter.*
  - *I have learned that it is cold in winter.*

After Reading:

• What details did I learn about this topic?
  - *I have learned that there are many fun things to see in winter.*
  - *I have learned that there are many things you wear in winter.*

• Read the book again and look for the vocabulary words.
  - *I see the word __scarf__ on page 8 and the word __skates__ on page 12. The other vocabulary words are found on page 14.*

What do you see in **winter**?

I see a blue **hat**.

I see a red hat.

I see a **scarf**.

I see **boots**.

I see **skates**.

Word List

Sight Words

a	in	what
blue	red	you
I	see	

Words to Know

boots

hat

scarf

skates

winter

26 Words

What do you see in **winter**?

I see a blue **hat**.

I see a red hat.

I see a **scarf**.

I see **boots**.

I see **skates**.

Seasons in a Year

WINTER

Written by: Amy Culliford

Designed by: Rhea Wallace

Series Development: James Earley

Proofreader: Kathy Middleton

Educational Consultant: Christina Lemke M.Ed.

Photographs:
Shutterstock: Smit: cover; Nemeziya: p. 1; Ryan DeBerardinis: p. 3, 14; Anna Klepatckaya: p. 5, 14; David Franklin: p. 6; Studio KIWI: p. 8-9, 14; NYS: p. 10, 14; monticello: p. 12-13, 14

Library and Archives Canada Cataloguing in Publication

Title: Winter / Amy Culliford.
Names: Culliford, Amy, 1992- author.
Description: Series statement: Seasons in a year | "A Crabtree roots book".
Identifiers: Canadiana (print) 20200386999 |
    Canadiana (ebook) 20200387014 |
    ISBN 9781427134752 (hardcover) |
    ISBN 9781427132727 (softcover) |
    ISBN 9781427132765 (HTML) |
    ISBN 9781427133144 (read-along ebook)
Subjects: LCSH: Winter—Juvenile literature.
Classification: LCC QB637.8 .C85 2021 | DDC j508.2—dc23

Library of Congress Cataloging-in-Publication Data

Names: Culliford, Amy, 1992- author.
Title: Winter / Amy Culliford.
Description: New York : Crabtree Publishing Company, 2021. | Series: Seasons in a year : a Crabtree roots book | Audience: Ages 4-6 | Audience: Grades K-1 | Summary: "Early readers are introduced to the winter season. Simple sentences and engaging pictures help describe what happens in winter"-- Provided by publisher.
Identifiers: LCCN 2020049776 (print) |
    LCCN 2020049777 (ebook) |
    ISBN 9781427134752 (hardcover) |
    ISBN 9781427132727 (paperback) |
    ISBN 9781427132765 (ebook) |
    ISBN 9781427133144 (epub)
Subjects: LCSH: Winter--Juvenile literature.
Classification: LCC QB637.8 .C85 2021 (print) | LCC QB637.8 (ebook) | DDC 508.2--dc23
LC record available at https://lccn.loc.gov/2020049776
LC ebook record available at https://lccn.loc.gov/2020049777

Crabtree Publishing Company

www.crabtreebooks.com   1-800-387-7650

Printed in Canada/012021/CPC20210118

Published in the United States
Crabtree Publishing
347 Fifth Avenue, Suite 1402-145
New York, NY, 10016

Published in Canada
Crabtree Publishing
616 Welland Ave.
St. Catharines, Ontario L2M 5V6